Questions in
Credit
Maths

Ken Nisbet

Text © 2004 Ken Nisbet
Design and layout © 2004 Leckie & Leckie Ltd
Cover image © Mehau Kulyk/Science Photo Library

04/120508

All rights reserved. No part of this publication may be reproduced, stored in a retrieval system, or transmitted in any form or by any means, electronic, mechanical, photocopying, recording or otherwise, without prior permission in writing from Leckie & Leckie Ltd. Legal action will be taken by Leckie & Leckie Ltd against any infringement of our copyright.

ISBN 978-1-84372-147-5

Published by
Leckie & Leckie Ltd, 3rd floor, 4 Queen Street, Edinburgh, EH2 1JE
Tel: 0131 220 6831 Fax: 0131 225 9987
enquiries@leckieandleckie.co.uk   www.leckieandleckie.co.uk

Special thanks to
David Collins (proofreading), Roda Morrison (copy-editing), River Design (design and cover)
and Robin Waterston (content review).

A CIP Catalogue record for this book is available from the British Library.

Leckie & Leckie Ltd is a division of Huveaux plc.

Contents

Introduction — 4

Skills and Techniques Revision — 5

1 Number and Money — 6
  1.1 Numbers and Accuracy — 6
  1.2 Calculations with Fractions — 6
  1.3 Calculations with Integers — 7
  1.4 Calculations with Indices — 7
  1.5 Calculations with Money — 8

2 Shape and Measure — 11
  2.1 Solids — 11
  2.2 Similarity — 12
  2.3 Right-angled Triangles — 13
  2.4 Circles — 15

3 Trigonometry — 18
  3.1 Sine, Cosine and Tangent — 18
  3.2 Three-figure Bearings — 19
  3.3 Solving Triangles — 19
  3.4 Trig Graphs — 24
  3.5 Trig Equations and Formulae — 25

4 Algebraic Relationships — 27
  4.1 Basic Operations and Notation — 27
  4.2 Brackets — 27
  4.3 Factorising — 28
  4.4 Solving Linear Equations — 29
  4.5 Algebraic Fractions — 31
  4.6 Surds and Indices — 32
  4.7 Solving Quadratic Equations — 34
  4.8 Variation — 35

5 Graphical Relationships — 37
  5.1 Gradient — 37
  5.2 Linear Graphs — 37
  5.3 Solving Problems Graphically — 39
  5.4 Quadratic Graphs — 40
  5.5 Other Types of Graphs — 42

6 Graphs, Statistics and Probability — 43
  6.1 Displaying Data — 43
  6.2 Statistics — 44
  6.3 Probability — 45

Exam Structure and Formulae — 46

Practice Exams — 47

Practice Exam A — 48
  Paper 1 — 48
  Paper 2 — 50

Practice Exam B — 52
  Paper 1 — 52
  Paper 2 — 54

Introduction

What this book contains

There are four sections in this question book.

- **Skills and Techniques Revision**

    This is a bank of questions designed to develop your skills and techniques in maths to a level where you will achieve success in examination questions. The skills and techniques concentrated on are those that occur most frequently in the exam.

- **Practice Exams**

    Two complete exams are provided. They have a similar content, structure and difficulty range to the exam you will be sitting.

- **Hints Section**

    Perhaps your approach to a question may not be working or your answer may be wrong or you may not even be sure where to start on a question. If you see this symbol beside a question, this indicates that a hint has been provided. Look in pages 15 to 20 of the pocket section and you will find useful guidance that will allow you to proceed to the solution.

- **Answers and Solutions**

    This section contains all the answers to the Skills and Techniques Revision Exercises and also the full solutions to the two Practice Exams.

Leckie & Leckie's Credit Level Maths Revision Notes

Questions in Credit Maths is the companion volume to Leckie & Leckie's *Credit Level Maths Revision Notes*. We recommend that you work with both books when revising for your exam. You will find extensive cross-references between the illustrative worked examples in the *Revision Notes* and the corresponding Revision Exercise in the *Questions*. Where you see the following book symbol beside a question, this indicates the number of the worked example from the *Revision Notes*.

Suggestions for using this book

There are many ways of using the revision material in this book. We recommend the following focused system:

Step 1: Sit Practice Exam A or Practice Exam B as a timed exercise.

Step 2: Compare your attempted solutions with the full solutions provided to identify your weak areas.

Step 3: Use the Skills and Techniques Revision Exercises to strengthen those weak areas.

Step 4: After completing Steps 1–3 wait a few days and then repeat the process. It is this repetitive practice of problems that will improve your chances of achieving a good grade in the final exam.

Additional comments

There is a wide range of difficulty in the Revision Exercises. If you devote enough time and effort to these exercises you will eventually master the skills and techniques that you need. A lot depends on your own belief that you can eventually succeed. Practice does make perfect. It is not possible in a book of this nature to cover all the material that is likely to occur in your exam and it is therefore vital that you attempt as many of the actual Past Papers as you can. Best wishes for success in your Credit Maths Exam and we hope that our revision books will help you towards this success.

Skills and Techniques Revision

1 Number and Money

1.1 Numbers and Accuracy

Exercise 1.1 Rounding Measurements

1 The following are calculator displays. Round them to **i** 3 significant figures **ii** 2 significant figures
   - **a** 29·28479105
   - **b** 9·453268471
   - **c** 0·455827934
   - **d** 125·7298479
   - **e** 1853·179215
   - **f** 0·047984991

2 Round each of these measurements to **i** 2 significant figures **ii** 1 significant figure
   - **a** 8·53 cm
   - **b** 0·945 km
   - **c** 0·0155 g
   - **d** 452 miles
   - **e** 1950 m
   - **f** 89·76 years

1.2 Calculations with Fractions

Exercise 1.2a Equivalent Fractions

Simplify:

1. $\frac{21}{28}$
2. $\frac{15}{25}$
3. $\frac{12}{16}$
4. $\frac{56}{63}$
5. $\frac{18}{45}$
6. $\frac{33}{77}$
7. $\frac{36}{48}$

Exercise 1.2b Multiplying Fractions

Evaluate (no calculator):

1. $\frac{7}{9} \times \frac{3}{14}$
2. $1\frac{1}{2} \times \frac{4}{9}$
3. $2\frac{1}{3} \times 1\frac{1}{2}$
4. $4\frac{1}{2} \times 2\frac{2}{3}$
5. $1\frac{1}{5} \times 7\frac{1}{2}$
6. $3\frac{3}{4} \times \frac{2}{5}$

Exercise 1.2c Dividing Fractions

Evaluate (no calculator):

1. $3 \div \frac{1}{3}$
2. $\frac{2}{5} \div 2$
3. $\frac{2}{3} \div \frac{1}{6}$
4. $1\frac{1}{2} \div 2\frac{1}{2}$
5. $3\frac{2}{3} \div 1\frac{5}{6}$
6. $3\frac{3}{4} \div 2\frac{1}{2}$

Exercise 1.2d Adding and Subtracting Fractions

Calculate (no calculator):

1. $\frac{2}{3} + \frac{4}{7}$
2. $1\frac{1}{3} - \frac{3}{4}$
3. $2\frac{2}{5} + 1\frac{1}{3}$
4. $\frac{1}{2} + 1\frac{4}{5}$
5. $3\frac{2}{3} - \frac{5}{6}$
6. $\frac{6}{7} + 2\frac{1}{3}$
7. $1\frac{1}{2} + 2\frac{1}{3} - \frac{1}{4}$
8. $3\frac{3}{8} + 1\frac{1}{2} - 2\frac{3}{4}$
9. $2\frac{7}{10} - \frac{3}{5} - 1\frac{1}{2}$

Exercise 1.2e Order of Operations

Calculate (no calculator):

1. $2\frac{1}{2} - \frac{3}{4} \times \frac{2}{3}$
2. $\frac{3}{4} + \frac{2}{5} \times \frac{5}{6}$
3. $\frac{2}{3}$ of $1\frac{1}{2} + \frac{1}{3}$
4. $1\frac{1}{2} + \frac{3}{4}$ of $\frac{2}{3}$
5. $\frac{1}{2}(1\frac{1}{5} - \frac{2}{3})$
6. $\frac{3}{7}(\frac{1}{5} + 1\frac{2}{3})$

See pages 6–10 of Leckie & Leckie's Credit Maths Revision Notes

Number and Money 1

1.3 Calculations with Integers

Exercise 1.3a Adding and Subtracting Integers

1. The rule for these sequences is 'add the two previous terms to get the next term'. Find the next three terms (no calculator):

   a 1, –3, … b 5, –1, … c –4, 1, …

2. In each brick tower the number on a brick is the sum of the two numbers on the bricks below it. Complete the numbers on these towers (no calculator):

   a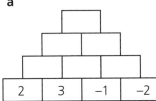
   b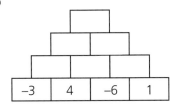
   c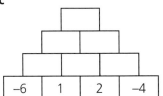

3. Calculate:

   a 2 – (–3) + 1   b –3 + (–2) – (–3)   c 1 – (–4) + (–2)
   d –4 + 3 – (–2)   e –1 – (–4) – (–3)   f –2 + (–3) – (–4)

Exercise 1.3b Multiplying and Dividing Integers

Calculate (no calculator):

1. $2 \times (-3)$
2. $-3 \times (-4)$
3. $(-5)^2$
4. $(-2)^3$
5. $\frac{-4}{2}$
6. $\frac{8}{-2}$
7. $\frac{(-3)^3}{9}$
8. $\frac{(-2)^4}{-8}$

Exercise 1.3c A Mixture

Calculate (no calculator):

1. $7 - 4 \times (-3)$
2. $(-2)^2 + 5 \times (-2)$
3. $\frac{3 - (-7)}{4 - (-1)}$
4. $8 \times (-3) - 3 \times (-3)^2$
5. $(-4)^2 - 2 \times (-4)$
6. $\frac{-2 - 5}{3 - (-11)}$
7. $(-2)^3 - 5 \times (-2)$
8. $7 \times (-1) - 3 \times (-1)^2$
9. $2 \times (-5)^2 - 3 \times (-5)$

1.4 Calculations with Indices

Exercise 1.4a Integral Indices

Evaluate (no calculator):

1. $3^0 + 2^{-1}$
2. $3^{-2} + 3^0$
3. $2^{-2} + 2^{-1}$
4. $\frac{1}{2}(2^{-1} + 2^0)$
5. $5^{-1}(2^2 + 2^0)$
6. $2^{-2}(2^0 - 2^{-1})$

1 Number and Money

1.4.2 Exercise 1.4b Fractional Indices

Evaluate (no calculator):

1. $27^{2/3}$
2. $81^{3/4}$
3. $64^{-2/3}$
4. $8^{-4/3}$
5. $81^{-3/4} \times 9^{3/2}$
6. $125^{-2/3} \times 25^{1/2}$
7. $100^{-3/2} \times 5^2$
8. $16^{-3/4} \times 4^{5/2}$

1.4.3 Exercise 1.4c Calculations in Scientific Notation

1. Calculate, giving your answer in scientific notation (no calculator):

    a. $(2 \times 10^3) \times (3 \times 10^5)$
    b. $(1\cdot 3 \times 10^{-6}) \times (2 \times 10^3)$
    c. $(8 \times 10^{12}) \times (2 \times 10^{-3})$
    d. $\dfrac{6 \times 10^{15}}{2 \times 10^3}$
    e. $\dfrac{8 \cdot 6 \times 10^4}{2 \times 10^{-12}}$
    f. $\dfrac{2 \times 10^{-3}}{4 \times 10^8}$

2. Use a calculator to do each calculation giving your answer in scientific notation to 3 significant figures where necessary.

    a. $7 \times 60 \times 60 \times (4 \times 10^7)$
    b. $(14 \times 10^{-6}) \times 16 \times 30$
    c. $4 \times 4 \times 20 \times (2 \cdot 36 \times 10^8)$
    d. $\dfrac{3 \cdot 1 \times 10^8}{60 \times 60}$
    e. $\dfrac{5 \cdot 7 \times 10^{-23}}{300}$
    f. $\dfrac{2 \cdot 38 \times 10^{17}}{5 \cdot 45 \times 10^{20}} \times 100$
    g. $3600 \times 60^2 \times 24 \times 365$

1.5 Calculations with Money

1.5.1 / 1.5.7 Exercise 1.5a Finding a % of a Quantity

1. VAT (Value Added Tax) is charged at 17·5%. Find, to the nearest penny, the VAT added to a bill of:

    a. £27
    b. £450
    c. £18·30
    d. £1847·50

2. The following prices increase in line with the given yearly inflation rate. Calculate the prices, 1 year on, to the nearest penny:

    a. Price: £23   Inflation Rate: $7\tfrac{1}{2}$%
    b. Price: £123   Inflation Rate: 4·5%
    c. Price: £2·63   Inflation Rate: $11\tfrac{1}{2}$%
    d. Price: £135   Inflation Rate: 12·5%

1.5.2 Exercise 1.5b Expressing one Quantity as a % of another Quantity

The following are price increases or decreases. In each case give the % increase or decrease giving your answer to 1 decimal place.

1. 49p to 51p
2. 89p to 83p
3. £2·60 to £2·70
4. £135 to £146
5. £260 to £235
7. £12 500 to £11 750

Number and Money 1

Exercise 1.5c Appreciation and Depreciation

1 Use a multiplier to:
   - **a** increase £230 by 15%
   - **b** decrease £1000 by 8%
   - **c** decrease £1250 by 3%
   - **d** increase £175 by 6%

2
   - **a** A house is worth £225 000 in 2006. If it appreciates in value by 8% per year, what is its value in 2009 to the nearest £1000?
   - **b** At 10 am there are 10 000 bacteria in a culture. If they increase at the rate of 12·5% per hour, how many are there by 1 pm (give your answer to three significant figures)?
   - **c** House contents are valued at £45 000 in 2005. If they depreciate in value by 5% per year, what is their value in 2008 to the nearest £?
   - **d** A patient at 9 am has 500 mg of a drug in her blood. If the amount of the drug reduces by 20% each hour, how many mg are in her blood at 12 noon?
   - **e** A marmoset colony is estimated at 2000 individuals in the year 2005. If the numbers decrease by 8% per year, estimate the size of the colony in 2009.

Exercise 1.5d Compound Interest

1 Calculate **i** the final amount **ii** the compound interest for the following investments:
   - **a** £640 for 3 years at 5% pa
   - **b** £2400 for 2 years at $2\frac{1}{2}$% pa

2
   - **a** A loan scheme has the following rules:
     - **(1)** Repay £200 on the 15th of each month
     - **(2)** Interest is charged at 1·5% per month on the amount outstanding at the end of each month.

     £2000 is borrowed on 1st February. Calculate the amount outstanding on 1st April.
   - **b** After two months how does this scheme compare with a scheme where interest is charged at 1·3% per month on the amount outstanding at the end of each month and where no repayments are made each month?

Exercise 1.5e Further Percentage Calculations

1 In each case calculate the original price before the increase or decrease:
   - **a** £265 after an increase of 6%
   - **b** £294·40 after a decrease of 8%
   - **c** £1344 after an increase of 12%
   - **d** £4300 after a decrease of 14%
   - **e** £30·08 after an increase of $17\frac{1}{2}$%
   - **f** £473·60 after a decrease of $7\frac{1}{2}$%

2
   - **a** The new 'Super Note Pad' has 870 pages, a 20% increase compared to the old 'Standard Note Pad'. How many pages has the 'Standard Note Pad'?
   - **b** A fridge sells for £260·85. This includes VAT at $17\frac{1}{2}$%. What is the price without VAT?
   - **c** A car is now valued at £11 900 having lost 15% of its value in a year. What was its value a year ago?

See Answers on pages 1–2 of answer booklet

1 Number and Money

Exercise 1.5f Ratio

1. Divide each of these totals into separate amounts according to the given ratio:

   a Total: £285   Ratio 1:4
   b Total: £136   Ratio 3:5
   c Total: £2530   Ratio 5:6
   d Total: £306   Ratio 1:2:3
   e Total: £23·60   Ratio 2:3:5
   f Total: £611   Ratio 1:5:7

2. Income from market stalls is divided between the Council, the Organisers and the Traders *in that order* in the ratio 1:2:7.

   a If the Traders receive £350 how much do the Organisers receive?

   b Income on Saturday was £4800. How much of this do the Traders receive?

3. Three employees at a café receive coins from the 'tip box' in the ratio 2:3 (pound coins to euro coins). How much does each receive if there are 35 one-pound coins and 45 one-euro coins in the 'tip box'?

   How much is left in the box after the distribution?

See pages 12–14 of Leckie & Leckie's Credit Maths Revision Notes

Shape and Measure 2

2.1 Solids

Exercise 2.1a Area Reminders

Calculate the area of each shape, giving your answer to 3 significant figures.

1
A semicircle

2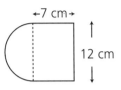
A rectangle and a semicircle

3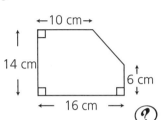

4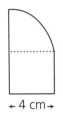
A square and a quarter circle

5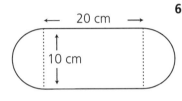
A rectangle and two semicircles

6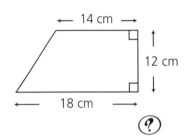

Exercise 2.1b Spheres and Cones

1 Find the volume of a sphere with radius:

   a 3 cm
   b 10·2 cm
   c 1·6 m
   d 43 cm

   (give your answers to 3 significant figures)

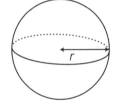

2 A slab of chocolate in the shape of a cuboid with dimensions 10 cm × 4 cm × 2 cm is melted and made into small spherical sweets with diameter 0·5 cm. How many sweets can be made?

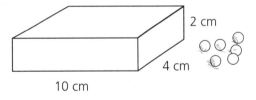

3 Find the volume of a cone with perpendicular height and base radius as given.

   a $h = 5$ cm, $r = 2$ cm
   b $h = 8·2$ cm, $r = 5$ cm
   c $h = 1·3$ m, $r = 0·8$ m
   d $h = 23$ cm, $r = 16$ cm

   (give your answers to 3 significant figures)

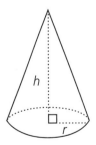

See Answers on page 2 of answer booklet

2 Shape and Measure

2.1.4 Exercise 2.1c   Prisms

For each of these prisms calculate its volume giving your answer to 3 significant figures.

1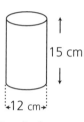

A cylinder

2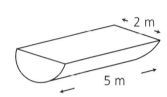

Semi-circular end

3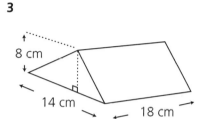

Triangular end with vertical height 8 cm

4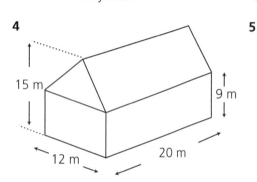

The end is in the shape of a rectangle surmounted by an isosceles triangle

5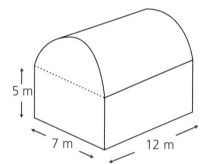

The end is in the shape of a rectangle surmounted by a semicircle

2.2 Similarity

2.2.1 Exercise 2.2a   Finding Lengths

1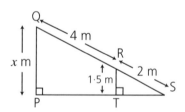

△PQS and △TRS are similar. Find x.

2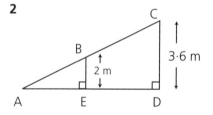

△ABE is similar to △ACD. If AC = 9 m, find the length of AB.

3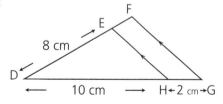

△DEH and △DFG are similar. Find the length of EF.

4

△MNQ is similar to △RPQ. Find the length of PR.

See pages 14–16 of Leckie & Leckie's Credit Maths Revision Notes

Shape and Measure 2

Exercise 2.2b Finding Areas and Volumes

1 In each case the two containers are mathematically similar.

a

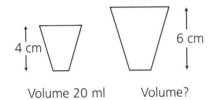

Volume 20 ml   Volume?

b

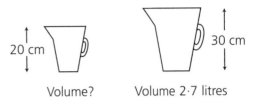

Volume?   Volume 2·7 litres

c

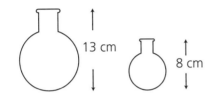

Volume = 2 litres

Find the volume of the smaller flask correct to the nearest ml.

d

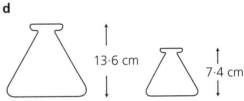

Volume = 250 ml

Find the volume of the larger flask in litres to 1 decimal place.

2 The small wooden cube has side length 4·5 cm. The large wooden cube has side length 8·5 cm. It takes 35 ml of varnish to paint the faces of one small cube. I have 1 litre of varnish. How many of the large cubes can I paint?

2.3 Right-angled Triangles

Exercise 2.3a Pythagoras' Theorem

1 In each case calculate x correct to 1 decimal place:

a   **b**

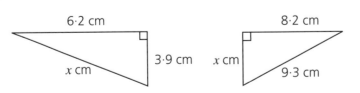

c

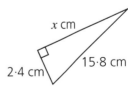

See Answers on page 2 of answer booklet

13

2 Shape and Measure

2 a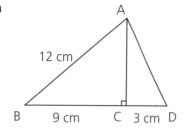

Calculate AD to 3 significant figures.

b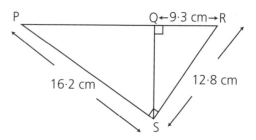

Calculate length PR to 1 decimal place.

3 A circular table of radius 1·5 m has a flap folded along AB as shown. Angle ACB is 90° where C is the centre of the table. Calculate the length AB to the nearest cm.

Exercise 2.3b The Converse of Pythagoras' Theorem

1 Determine in each case whether or not the triangle is right-angled:

a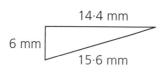

b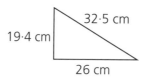

c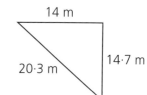

2 Determine whether each picture frame is truly rectangular:

a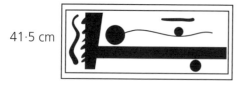

Diagonal measures 107·5 cm

b

Diagonal measures 146·5 cm

c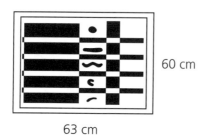

Diagonal measures 88 cm

See pages 16–18 of Leckie & Leckie's Credit Maths Revision Notes

Shape and Measure 2

2.4 Circles

Exercise 2.4a Finding Arcs and Sectors

1 For each sector find its **i** perimeter **ii** area, giving your answers correct to 3 significant figures.

a   b   c  d

2 a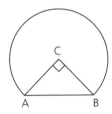

A circular table with radius 75 cm has a flap folded along AB as shown. Angle ACB = 90° where C is the centre of the table. Calculate the perimeter of the table to the nearest mm.

b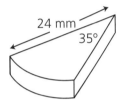

A wooden playing piece is in the shape of a prism whose end is the sector of a circle with radius 24 mm and sector angle 35°. Find:
i   the area of the sector
ii the volume of the piece if it is 4 mm thick
(give both answers correct to 3 significant figures)

Exercise 2.4b Find the Angle at the Centre

1 a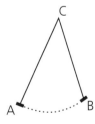

A 3 m long swing is attached to a frame at C as shown. The seat moves along arc AB which is 2·5 m long. Find the angle through which the swing turns going from positions CA to CB.

b

The minute hand of this clock is 17·5 cm long. Through what angle has the hand turned if its tip travels 10 cm?

2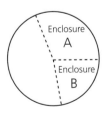

A circular field with radius 60 m has two sectors, enclosure A and enclosure B, fenced off as shown. Enclosure A has area 3100 m² and enclosure B has area 2500 m². Find, to 1 decimal place, the angle at the centre of the field made by each enclosure.

2 Shape and Measure

2.4.3 Exercise 2.4c Circle Tangents

1 In each of these Circle/Tangent diagrams find the value of x to 3 significant figures.

a 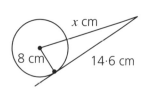 b 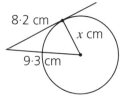 c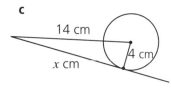

2 a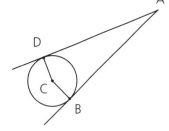

Two tangents to a circle, centre C, are drawn from A touching the circle at B and D as shown. If AC = 19 cm and the circle has radius 5 cm calculate the perimeter of kite ABCD correct to 3 significant figures.

b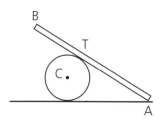

A stick AB rests on a circular barrel, centre C radius 90 cm, touching the barrel at T and the ground at A. AC = 3·6 m and the overhang BT = 1 m. Calculate the length of the stick correct to the nearest cm.

2.4.4 Exercise 2.4d Angles in Semi-circles

1 In each of these semicircle diagrams calculate x to 3 significant figures:

a 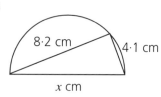 b 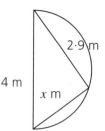 c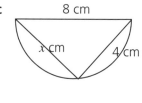

2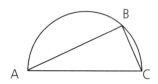

a Two supports AB and BC in a semicircular tunnel are 3 m and 2 m in length. How wide is the tunnel to the nearest cm?

b Four posts A, B, C and D are placed on the circumference of a circular field. AC is a diameter.
  i AB = 8 m and BC = 3 m. Calculate the diameter to the nearest cm.
  ii If AD = 7 m calculate DC to the nearest cm.

 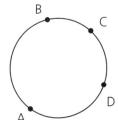

See pages 18–19 of Leckie & Leckie's Credit Maths Revision Notes

Shape and Measure 2

Exercise 2.4e Symmetry and Chords

1 a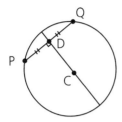

radius is 8 cm DC = 6 cm
Calculate PQ in cm to 1 decimal place.

b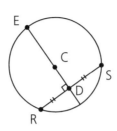

radius is 2 m ED = 3.5 m
Calculate RS to 3 significant figures.

c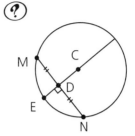

radius is 9.3 cm ED = 2 cm
Calculate MN to the nearest mm.

2 a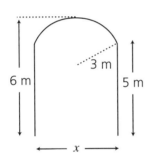

The diagram shows one section of a bridge with a circular arch of radius 3 m. Calculate x, the width of the section, to the nearest cm.

b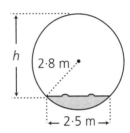

This circular rail tunnel has radius 2.8 m. The width of the rail bed is 2.5 m. Calculate h, the height of the tunnel, to the nearest cm.

c

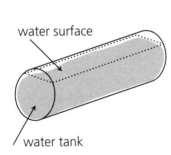 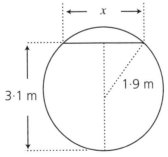

This cross-section of the water tank shows a radius of 1.9 m and maximum depth of water at 3.1 m. Find x, the width of the water surface, to the nearest cm.

See Answers on page 2 of answer booklet

3 Trigonometry

3.1 Sine, Cosine and Tangent

Exercise 3.1a Finding Sides and Angles

1 For each triangle calculate *x* correct to 3 significant figures:

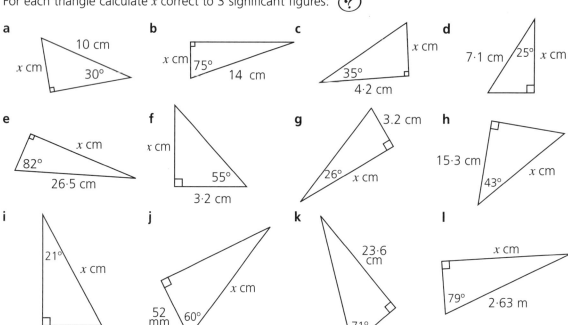

2 In each case find the value of *x* correct to 1 decimal place:

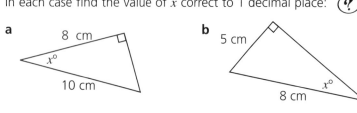

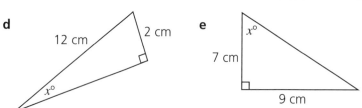

3

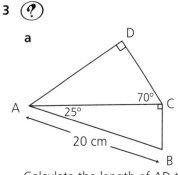

Calculate the length of AD to 3 significant figures.

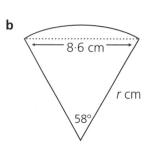

Calculate the value of the radius, *r*, of this circle segment (answer to 1 decimal place).

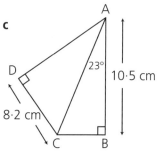

Calculate the size of angle CAD correct to 1 decimal place.

See pages *20–22* of Leckie & Leckie's Credit Maths Revision Notes

Trigonometry 3

Exercise 3.1b Angles greater than 90°

Evaluate the following to 3 significant figures:

1. sin 240°
2. cos 130°
3. tan 269°
4. 1 + sin 100°
5. tan 325° − 1
6. sin 23° + sin 123°
7. −2 − tan 289°
8. cos 1000°
9. $\sin^2 100° + \cos^2 100°$ (repeat with a different angle, e.g. $\sin^2 252° + \cos^2 252°$, etc)
10. $\frac{\sin 193°}{\cos 193°}$ (compare with tan 193°. Repeat with different angles.)

3.2 Three-figure Bearings

Exercise 3.2 Three-figure Bearings

1. Draw diagrams to illustrate the following situations and to help you calculate the required bearing:

   a. B is on a bearing of 100° from A. Find the bearing of A from B.

   b. H is on a bearing of 025° from G. Find the bearing of G from H.

   c. P is on a bearing of 338° from Q. Find the bearing of Q from P.

2. a. From B, A is on a bearing of 050° and C is on a bearing of 125°. Find the size of angle ABC.

   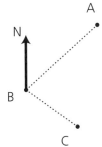

   b. From airport A an aircraft flies towards P on a bearing of 063° and another aircraft flies towards Q on a bearing of 106°. Find the size of angle PAQ.

   c. From lighthouse L the bearing of ship A is 045°. From ship A the bearing of ship B is 200°. Find the size of angle LAB.

3. For each situation i Draw a diagram to illustrate the information.
                      ii Find the sizes of the three angles of triangle ABC.

   a. From A, B is on a bearing of 140° and C is on a bearing of 200°. Also from C, B is on a bearing of 080°.

   b. From A, B is on a bearing of 290° and C is on a bearing of 030°. Also from B, C is on a bearing of 070°.

   c. From C, A is on a bearing of 225°. From A, B is on a bearing of 110°. From B, C is on a bearing of 350°.

3.3 Solving Triangles

Exercise 3.3a Area of a Triangle

1. Find the area of each triangle. Give your answers to 3 significant figures.

   a 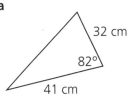 b 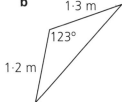 c 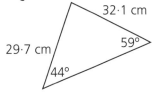 d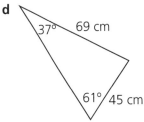

3 Trigonometry

2 a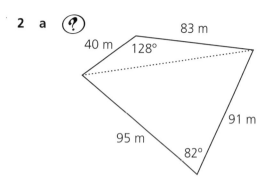

The diagram shows the plan of a fenced enclosure. Find the area of the enclosure to 3 significant figures.

b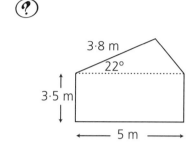

The diagram shows the rectangular end of a garden hut surmounted by a triangle. The end is to be painted. Find the area to be painted to the nearest square metre.

3 a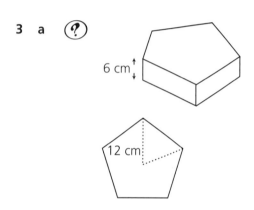

This gift box is in the shape of a pentagonal prism and is 6 cm deep. The box has a uniform cross-section that is a rectangular pentagon as shown with centre to vertex 12 cm. Calculate the volume of the box to the nearest cm³.

b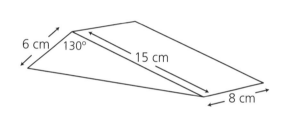

This wooden wedge is in the shape of a triangular prism. It has a uniform cross-section with dimensions as shown in the diagram. Calculate, to 3 significant figures, the volume of this wedge.

4 a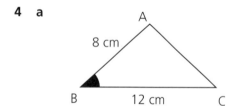

The area of triangle ABC is 35 cm². Calculate the size of the acute angle ABC to 1 decimal place.

b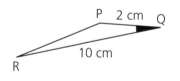

The area of triangle PQR is 2 cm². Calculate the size of the acute angle PQR to 1 decimal place.

c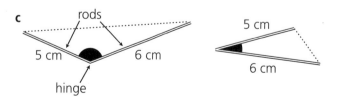

As shown in the diagrams there are two different positions that the two rods can be in to form a triangle with area 5 cm². Calculate the two possible sizes of the angle between the rods to 1 decimal place.

See page 22 of Leckie & Leckie's Credit Maths Revision Notes

d Triangle ABC has area of 16 cm². AB is 4 cm and AC is 12 cm. Calculate possible sizes of angle BAC to 1 decimal place.

e Triangle LMN has area of 20 cm². ML is 8·2 cm and MN is 6·3 cm. Calculate possible sizes of angle LMN to 1 decimal place.

Exercise 3.3b The Sine Rule – finding sides

1 a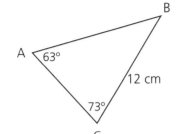

Calculate AB
(to 3 significant figures)

b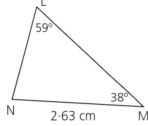

Calculate LN
(to 3 significant figures)

c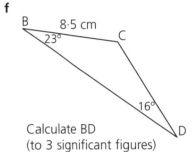

Calculate DE
(to 3 significant figures)

d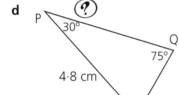

Calculate PQ
(to 3 significant figures)

e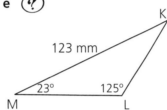

Calculate ML
(to 3 significant figures)

f

Calculate BD
(to 3 significant figures)

2 a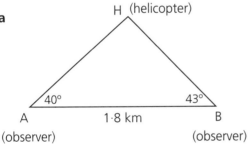

Two observers, 1·8 km apart, are tracking a helicopter. From observer A the helicopter has an angle of elevation of 40° and from B an angle of elevation of 43°. How far is the helicopter from observer B?

b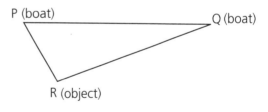

Two boats, 750 m apart, detect an object on their sonar screens lying on the ocean bed. From one boat the angle of depression of the object is 60° and from the other it is 18°. How far is the object from each of the boats?

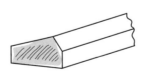 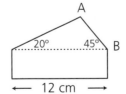

c A wooden rod has a uniform cross-section in the shape of a rectangle surmounted by a triangle as shown in the diagram. Calculate the length AB to 3 significant figures.

3 Trigonometry

3 **a**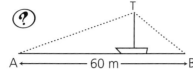

Boat A and Boat B are 60 m apart. From A the angle of elevation of the top of the mast, T, of a yacht is 22°. From B the angle of elevation of T is 34°. Calculate the height of T above sea-level in metres to 1 decimal place.

b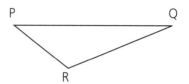

Boat P and Boat Q are 3 km apart. Object R is detected at an angle of depression of 36° from P and angle of depression 25° from Q. Calculate the depth of R in km to 3 significant figures.

c Two bird watchers are 150 m apart. They see a buzzard. From one of the watchers the bird has an angle of elevation of 49° and from the other the angle of elevation is 71°. If the bird is in between the two watchers calculate its height to the nearest metre.

Exercise 3.3c   The Sine Rule – finding angles

1 **a** Calculate angle P (to 1 decimal place)

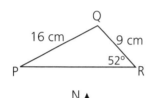

b Calculate angle B (to 1 decimal place)

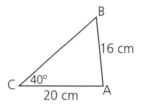

c Calculate angle F (to 1 decimal place)

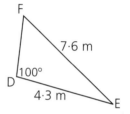

2 **a**

B is due South of A. The bearing of C from B is 140°. Calculate the bearing of C from A.

b

P is due North of Q. R is on a bearing of 255° from Q. Calculate the bearing of R from P.

c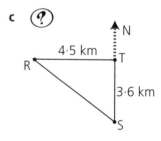

S is due South of T. R is on a bearing of 305° from S. Calculate the bearing of R from T.

Exercise 3.3d   The Cosine Rule – finding sides

1 **a**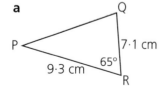

Calculate PQ (to 3 significant figures)

b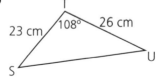

Calculate SU (to 3 significant figures)

c

Calculate DF (to 3 significant figures)

Trigonometry 3

2 a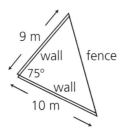

The diagram shows a triangular enclosure with two walls at an angle of 75°. Calculate the length of the fence to the nearest cm.

b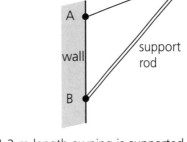

A 1·2 m length awning is supported by a rod. The awning makes an angle of 110° with the wall. The distance AB between where the awning and rod are attached to the wall is 1 m. Calculate the length of the rod to the nearest cm.

c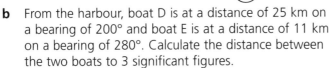

A golfer hits his ball 198 m towards the hole but 15° away from the line from the tee to the hole, a distance of 230 m. Calculate the distance of the ball from the hole to the nearest metre.

3 a Boat A is on a bearing of 065° from the harbour. Boat B, again from the harbour, is on a bearing of 110°. Calculate, using the diagram, the distance between the boats to 1 decimal place (in km). (?)

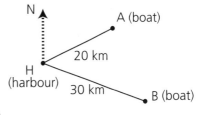

b From the harbour, boat D is at a distance of 25 km on a bearing of 200° and boat E is at a distance of 11 km on a bearing of 280°. Calculate the distance between the two boats to 3 significant figures.

c From boat P, boat Q is at a distance of 180 km on a bearing of 063°. A rock lies 230 km from Q on a bearing of 223°. How far is the rock from boat P to the nearest km? (?)

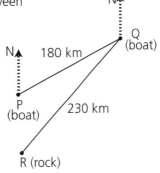

Exercise 3.3e The Cosine Rule – finding angles

1 a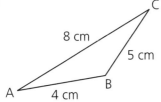

Calculate the size of angle BAC to 1 decimal place.

b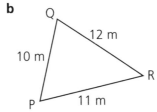

Calculate the size of angle QRP to 1 decimal place.

c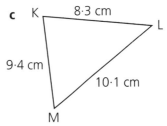

Calculate the size of angle MKL to 1 decimal place.

3 Trigonometry

2 a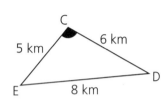

Calculate the size of angle BCD to 1 decimal place.

b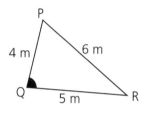

Calculate the size of angle LMN to 1 decimal place.

c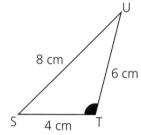

Calculate the size of angle STU to 1 decimal place.

3 a (non-calculator)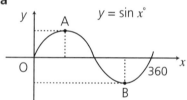

Show that $\cos C = -\frac{1}{20}$

b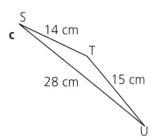

Show that $\cos Q = \frac{1}{8}$

c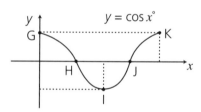

Show that $\cos T = -\frac{1}{4}$

3.4 Trig Graphs

Exercise 3.4a The Sine, Cosine and Tangent Graphs

1 For $0 \leq x \leq 360$, on the same diagram sketch the graphs of:

a $y = \sin x°$ and $y = \tan x°$ **b** $y = \cos x°$ and $y = \tan x°$

c $y = \sin x°$, $y = \cos x°$ and $y = \tan x°$

2 a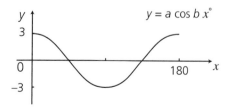

Write down the coordinates of A and B.

b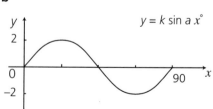

Write down the coordinates of G, H, I, J and K.

Exercise 3.4b Related Graphs

1 a

$y = a \cos b x°$

Find the values of a and b.

b

$y = k \sin a x°$

Find the values of k and a.

Trigonometry 3

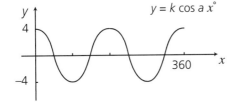

c $y = k \cos a x°$

Find the values of k and a.

d $y = a \sin b x°$

Find the values of a and b.

2 a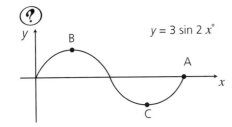

$y = 3 \sin 2 x°$

Write down the coordinates of A, B and C.

b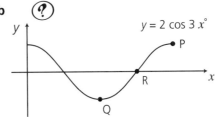

$y = 2 \cos 3 x°$

Write down the coordinates of P, Q and R.

3.5 Trig Equations and Formulae

Exercise 3.5a Solving Simple Trig Equations

1 Solve these equations for $0 \leq x \leq 360$
   a $\sin x° = 1$
   b $\cos x° = -1$
   c $\cos x° = 1$
   d $\tan x° = 0$
   e $\sin x° = 0$
   f $\sin x° = -1$

2 Solve these equations algebraically for $0 \leq x < 360$
   a $9 \cos x° - 1 = 0$
   b $3 \sin x° - 2 = 0$
   c $7 \sin x° + 2 = 0$
   d $1 + 3 \tan x° = 0$
   e $2 + 7 \cos x° = 0$
   f $5 - 7 \sin x° = 0$
   g $3 \sin x° + 1 = -1$
   h $2 + 5 \cos x° = 1$
   i $-3 = 9 \tan x° + 1$
   j $\cos 50° = 3 \cos x° + 1$
   k $\sin 20° = 4 \sin x° + 1$
   l $\tan 35° = 2 \tan x° + 3$

3 a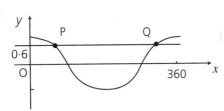

The diagram shows part of the graph of $y = \cos x°$. The line $y = 0.6$ cuts the graph at P and Q. Find the x-coordinates of P and Q.

b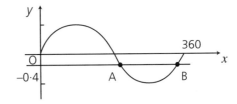

The diagram shows part of the graph of $y = \sin x°$. The line $y = -0.4$ cuts the graph at A and B. Find the x-coordinates of A and B.

Exercise 3.5b Trig Formulae

1 Show that $\frac{1 - \cos^2 A}{\cos^2 A} = \tan^2 A$

2 Show that $\cos^2 x - \sin^2 x = 2 \cos^2 x - 1$

3 Show that $1 - 2 \sin^2 x = \cos^2 x - \sin^2 x$

4 Show that $\frac{1 - \cos^2 y}{1 - \sin^2 y} = \tan^2 y$

3 Trigonometry

Exercise 3.5c Applications of Trig Formulae

1. The depth of water, D metres, at a harbour, t hours after noon is given by the formula:
   $D = 14{\cdot}6 + 7{\cdot}9 \sin (30t)°$

   a Find the maximum and minimum depths of the water during the course of a day.

   b Find the depth of the water at 1am.

   c When is the depth of the water 20 metres for the first time after noon?

2.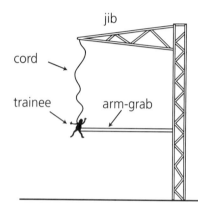

   A trainee bungee jumper is suspended on a cord from a jib. An arm-grab holds her preventing her from falling. When released, her height, h metres, above the ground is given by:

   $h = 16 + 10 \cos (36t)°$

   where t is the time in seconds after her release.

   a Find her maximum and minimum heights after she is released.

   b How long does it take her to first return to her starting height?

   c How high off the ground is she after 3 seconds?

   d How long does it take her to fall the first 10 metres?

   e When is her height 15 metres for the first time?

3. The diagram shows a design for an electricity generator on a wind farm. The height, h metres, of the tip of the blade (A) above the ground is given by:
   $h = 9{\cdot}5 + 5{\cdot}5 \cos (72t)°$ where t is the number of seconds after the release of the blade.

   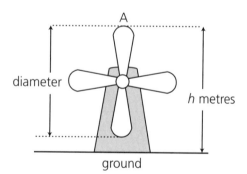

   a Find the diameter of the rotor blades (as shown on diagram).

   b Find the height of point A after 2·6 seconds.

   c When is point A first at a height of 12 metres?

Algebraic Relationships 4

4.1 Basic Operations and Notation

Exercise 4.1a Adding, Subtracting and Multiplying Terms

1 Simplify:

   a $a - (-2a)$
   b $a \times (-2a)$
   c $(-2a)^2$
   d $-3x^2 - x + x^2 - 2x$
   e $y^2 + y \times (-2y)$
   f $-3ab - a \times (-b)$
   g $(-a)^2 - (-a) \times 2a + a \times (-3a)$
   h $-3 - 2x - (-2)^2 - (-x) + 7$

2 a $Q = 3a^2 - b^2$ Calculate Q when

   i $a = 2$ and $b = -3$
   ii $a = -4$ and $b = -1$
   iii $a = -1$ and $b = -2$

   b $R = x^2 + 3y^2$ Calculate R when

   i $x = -1$ and $y = 2$
   ii $x = -2$ and $y = -3$
   iii $x = -\frac{1}{2}$ and $y = -\frac{1}{2}$

   c $A = k^2 - 2m^2$ Calculate A when

   i $k = -2$ and $m = -1$
   ii $k = -1$ and $m = 3$
   iii $k = -5$ and $m = -3$

Exercise 4.1b Function Notation

1 a $g(x) = 6 - 3x$ Evaluate:

   i $g(-1)$
   ii $g(3)$
   iii $g(-2)$
   iv $g(-\frac{1}{2})$

   b $h(x) = 2x - x^2$ Evaluate:

   i $h(2)$
   ii $h(-2)$
   iii $h(-5)$
   iv $h(-\frac{1}{2})$

   c $f(n) = n^2 - 4n$ Evaluate:

   i $f(-1)$
   ii $f(3)$
   iii $f(-2)$
   iv $f(-\frac{1}{2})$

2 a Given that $f(x) = 5 - 3x$ and that $f(m) = -1$, find m. (?)
   b Given that $g(x) = \frac{2x+1}{3}$ and that $g(n) = -3$, find n. (?)
   c $h(x) = 2 + 3x^2$. Given that $h(k) = 29$ find k if $k \geq 0$. (?)
   d $f(x) = -7x + 2$. Given that $f(n) = -12$, find n. (?)

3 In each of the following, two functions are defined. Find the values of x for which $f(x) = g(x)$.

   a $f(x) = x^2$ and $g(x) = 3x$
   b $f(x) = 2x^2$ and $g(x) = 4x$
   c $f(x) = x^2$ and $g(x) = -5x$
   d $f(x) = x^2 + x$ and $g(x) = 5x$
   e $f(x) = x^2 - 3x$ and $g(x) = 2x$
   f $f(x) = x - x^2$ and $g(x) = -x$
   g $f(x) = \frac{9}{x}$ and $g(x) = x$
   h $f(x) = \frac{16}{x}$ and $g(x) = 4x$

4.2 Brackets

Exercise 4.2a One Pair of Brackets

1 Simplify:

   a $4(x - 3)$
   b $-3(y + 2)$
   c $-(a - b)$
   d $-5(2 - x)$
   e $6(x^2 - 2x)$
   f $-2(3x - 5x^2)$

4 Algebraic Relationships

2 Simplify by removing brackets and collecting like terms:

a $4(2x - 3) - 5(x + 2)$
b $-2(3a - 2b) + 3(a - 5b)$
c $-(3x^2 - x) - 2(x - 2x^2)$
d $-3(2y - y^2) - (3y - 2y^2)$

Exercise 4.2b Two pairs of brackets

Remove brackets and collect like terms:

1 $(x - y)(x + 2y)$
2 $(2a - b)(a + 3b)$
3 $(m + 2n)(m - n)$
4 $(4x - y)(3x - 2y)$
5 $(3p - 2q)(4p + q)$
6 $(3x - 2y)(4x - 3y)$
7 $(2r + 3s)(r - s)$
8 $(5e - 3f)(2e - 6f)$
9 $(5x - 6y)(5x + 6y)$

Exercise 4.2c Brackets Squared

Remove brackets and and simplify where possible:

1 $(2a - 3b)^2$
2 $(x + 3y)^2$
3 $(5m - 6n)^2$
4 $(2p + 7q)^2$
5 $(8k - 3g)^2$
6 $(a - 10b)^2$
7 $(a - b)^2 + (a + b)^2$
8 $(2x - y)^2 - (x + 2y)^2$
9 $(5x - 2y)^2 - 2(x + y)^2$
10 $3(4 - a)^2 - 2(3 - 4a)^2$

Exercise 4.2d Trinomials and Brackets

Multiply out and collect like terms:

1 $(y - 3)(y^2 + 2y - 4)$
2 $(x + 4)(x^2 - x - 3)$
3 $(2m + 1)(m^2 - 3m + 2)$
4 $(3w - 1)(w^2 + 2w + 7)$
5 $(3 - x)(2 - 3x + x^2)$
6 $(7 - 2x)(1 + x - 3x^2)$

4.3 Factorising

Exercise 4.3a Common Factors

1 Factorise completely:

a $3x^2 - 9x$
b $10b - 15b^2$
c $4ab + b^2$
d $8a^2 + 12ab$
e $4ab^2 + 2a^2b$
f $36m^2 - 8mn$
g $2ab - 4ac + 6ad$
h $3x - 6x^2 + 9xy$
i $6ab^2 + 3a^2b - 9ab$

2 a A sequence like 2, 5, 7, 12, 19, 31, ... where each term after the first two is the sum of the previous two terms, is called a Fibonacci Sequence. Show that in *any* Fibonacci Sequence (no matter what the first two terms are):

   i   The sum of the 1st and 5th terms is a multiple of 3.
   ii The sum of the first six terms is a multiple of 4.
   iii The sum of the 2nd and 6th terms is three times the 4th term.

  b The number 47 has value $10 \times 4 + 7$. In general the two digit number '*ab*' has value $10a + b$.

   i   What value has the two digit number '*ba*'?
   ii Show that any two digit number added to the same number backwards is a multiple of 11. For example $47 + 74 = 121$, a multiple of 11.

Algebraic Relationships 4

Exercise 4.3b Differences of Two Squares

1 Factorise:
   a $4b^2 - 25c^2$
   b $x^2 - 64$
   c $81 - a^2$
   d $9p^2 - 16q^2$
   e $36a^2 - 100$
   f $49m^2 - 121n^2$
   g $1 - 25w^2$
   h $144x^2 - 1$
   i $81y^2 - 400$

2 Factorise fully:
   a $2x^2 - 2$
   b $3y^2 - 27$
   c $16x^2 - 4y^2$
   d $45a^2 - 80b^2$
   e $72m^2 - 8n^2$
   f $7k^2 - 175m^2$
   g $162a - 8ax^2$
   h $27xa^2 - 12xb^2$
   i $80w^2 - 500z^2$

Exercise 4.3c Quadratic Expressions

1 Factorise:
   a $x^2 + 3x - 10$
   b $a^2 + 5a - 36$
   c $y^2 - 2y - 24$
   d $n^2 - n - 12$
   e $x^2 - 9x + 18$
   f $w^2 - 10w + 16$
   g $x^2 + 2x - 48$
   h $a^2 - a - 72$
   i $k^2 + 2k - 48$

2 Factorise:
   a $2n^2 + 5n + 3$
   b $10m^2 + 19m + 6$
   c $2x^2 - 7x + 3$
   d $2y^2 - 5y + 2$
   e $3a^2 - 2a - 8$
   f $3x^2 - 5x - 2$
   g $6a^2 - 13a + 6$
   h $12x^2 - 11x - 5$
   i $6w^2 + 5w - 6$

3 Factorise fully:
   a $2x^2 + 8x - 42$
   b $3x^2 - 21x - 24$
   c $10x^2 - 10x - 420$
   d $3a^2 + 42a - 216$
   e $5w^2 - 35w + 60$
   f $18x^2 + 12x + 2$
   g $12y^2 + 21y - 6$
   h $a^4 - 81$
   i $2x^4 - 32y^4$

4.4 Solving Linear Equations

Exercise 4.4a Solving Linear Equations

1 Solve:
   a $10 - 3(2x + 1) = x$
   b $2x - 2(3 - 2x) = 3$
   c $4(2 - x) - 5x = -4$
   d $11 = 14 - 3(2 - 5x)$
   e $-3(x - 3) = 2(5 - x)$
   f $1 + 4(2 - 3x) = -3(x + 6)$

2 Solve:
   a $\frac{2}{3}x = 4$
   b $\frac{1}{4}x = 3$
   c $\frac{2}{5}x = 1$
   d $\frac{3}{7}x = 2$

4 Algebraic Relationships

3 Solve algebraically:

a $\dfrac{x}{2} = \dfrac{5-x}{3}$
b $3x - \dfrac{(2x+1)}{2} = 1$
c $\dfrac{x+1}{3} - \dfrac{(x+2)}{5} = 1$
d $\dfrac{2x-1}{2} - \dfrac{4-3x}{3} = -2$

4 On a motorway, if the average speed of vehicles is v metres/sec then the number of cars, C, passing a census point during a 10 minute period is given by the formula: $C = \dfrac{500v}{5+v}$

Calculate the speed of the cars in metres per second if in a 10 minute period 250 cars pass the census point.

Exercise 4.4b Solving Linear Inequations

Solve algebraically the following inequalities:

1 $3 - 2x < 4x - 3$
2 $3x > 2 - (x - 6)$
3 $5x - 7 > 3(1 + 5x)$

4 $1 + 3x \geqslant 7x - 7$
5 $-2(1 + 3x) \leqslant 3x + 1$
6 $x - 2(3x + 1) > 4x$

Exercise 4.4c Simultaneous Equations

1 Solve algebraically the system of equations:

a $2x + 3y = 13$
$3x + 2y = 12$

b $4x - y = 19$
$3x + 4y = 19$

c $5m - 3n = 15$
$6m - 5n = 11$

d $2x + 3y = -7$
$3x - 5y = 18$

e $2p - 7q = 1$
$5p + 2q = -17$

f $5e - 3f = 7$
$9e - 7f = 15$

2 Set up a linear equation for each of these situations. Remember to state clearly the quantity represented by each letter you use.

a Three adults and four children go to the theatre. Their tickets cost £54 in total.

b The total cost of my journey was £18. The bus cost 25p per km and the taxi charged 65p per km.

3 Zoe and Mia buy fruit at the market.

a Zoe buys 2 apples and 3 oranges for £1·65. Write down an algebraic equation to illustrate this.

b Mia buys 5 apples and 4 oranges for £2·90. Write down an algebraic equation to illustrate this.

c By solving algebraically the system of equations you have written down, find the cost of 3 apples and 2 oranges.

4 Kim travels regularly from home to work either catching a bus or taking a taxi.

a 3 bus trips and 2 taxi trips cost her £12·50. Write down an algebraic equation to illustrate this.

b 2 bus trips and 5 taxi trips cost her £20·25. Write down an algebraic equation to illustrate this.

c By solving algebraically the system of equations find the cost of 4 bus trips and 3 taxi trips.

5 Theatre tickets cost £7·50 for the stalls and £10·50 for the circle.

a A group of 12 theatre goers split themselves, some sitting in the stalls and some sitting in the circle. Illustrate this information by writing down an equation.

b The total cost of the tickets for the group is £111. Write another equation using this information.

c By solving the system of equations determine how many sat in the stalls and how many sat in the circle.

See pages 34–39 of Leckie & Leckie's Credit Maths Revision Notes

Algebraic Relationships 4

6 In a Fibonacci Sequence, each term after the first two is the sum of the two previous terms. For example: −1, 5, 4, 9, 13, …

   a Here is a Fibonacci Sequence: $x, y, \_, \_, \_, 1$. The 1st term is x and the 2nd term is y. The 6th term is 1. Find an equation in x and y.

   b Swapping the 1st and 2nd terms results in the following Fibonacci Sequence: $y, x, \_, \_, \_, 23$. Write down another equation in x and y.

   c Find the values of x and y.

4.5 Algebraic Fractions

Exercise 4.5a Cancelling Fractions

Simplify:

1 $\dfrac{m^2 - n^2}{m - n}$
2 $\dfrac{6a^2 - 18a}{a^2 - 9}$
3 $\dfrac{2(3k - 2)}{9k^2 - 4}$
4 $\dfrac{9g^2 - 4e^2}{3g + 2e}$
5 $\dfrac{x^2 - 2x - 3}{x - 3}$

6 $\dfrac{12y - 6}{2y^2 + 5y - 3}$
7 $\dfrac{6a^2 - 6}{2a^2 + a - 3}$
8 $\dfrac{81 - m^2}{9 - 8m - m^2}$
9 $\dfrac{9x^2 - 16}{3x^2 - 2x - 8}$

Exercise 4.5b Multiplying Fractions

Simplify:

1 $\dfrac{3a}{4} \times \dfrac{8b}{ab}$
2 $\dfrac{4(a - b)}{5} \times \dfrac{10}{(a - b)^2}$
3 $\dfrac{3(2x + 1)}{2(2x - 1)} \times \dfrac{4(2x - 1)}{(2x + 1)^2}$

4 $\dfrac{x(x + 1)(x - 3)}{5} \times \dfrac{10}{x(x + 1)}$
5 $\dfrac{3a^2 b}{(a + 3)^2} \times \dfrac{a + 3}{3ab}$
6 $\dfrac{4m}{n(m - n)} \times \dfrac{(m + n)(m - n)}{2m}$

Exercise 4.5c Dividing Fractions

Simplify:

1 $6 \div \dfrac{3}{x}$
2 $\dfrac{2}{a} \div 4$
3 $\dfrac{a^2}{b} \div \dfrac{a}{b}$
4 $\dfrac{2m}{3} \div \dfrac{m^2}{6}$

5 $\dfrac{2x - 1}{3x} \div \dfrac{(2x - 1)^2}{x^2}$
6 $\dfrac{4(m - 1)}{3(m + 1)^2} \div \dfrac{2(m - 1)^2}{(m + 1)}$

Exercise 4.5d Adding and Subtracting Fractions

1 Express as a single fraction in simplest form:

   a $\dfrac{1}{3k} - \dfrac{1}{4k}$
   b $\dfrac{2a - 1}{2a} + \dfrac{2 - a}{4a}$
   c $\dfrac{2x + 3}{3x} - \dfrac{1}{x}$

   d $\dfrac{1}{a} + \dfrac{a - b}{ab}$
   e $\dfrac{3 + m}{3m} - \dfrac{1}{3}$
   f $\dfrac{3}{20} + \dfrac{10 - 3y}{20y}$

2 Simplify:

   a $\dfrac{2}{a + 1} + \dfrac{3}{a}$
   b $\dfrac{6}{x} - \dfrac{5}{x - 3}$
   c $\dfrac{1}{x - 1} - \dfrac{1}{x + 1}$

   d $\dfrac{3}{2y} - \dfrac{1}{y + 1}$
   e $\dfrac{1}{a - 2} - \dfrac{a + 4}{3a(a - 2)}$
   f $\dfrac{a + b}{b} - 1$

Exercise 4.5e Changing the Subject of a Formula

In each case change the subject of the formula to the letter indicated:

1 a $P = \tfrac{1}{3}(a - b)$ to a
   b $A = 2 - \tfrac{3}{B}$ to B
   c $C = \dfrac{m - n}{w}$ to m

   d $m = 5A + B^2$ to A
   e $x = \sqrt{\dfrac{y}{z}}$ to y
   f $Q = \dfrac{a(b + c)}{3}$ to a

   g $a = \sqrt{b^2 + c^2}$ to b
   h $\dfrac{a}{\sin A} = \dfrac{b}{\sin B}$ to $\sin B$

4 Algebraic Relationships

2 a This circular earring is made from 8 cm of wire. Show that the area of the circular hole is exactly $\frac{9}{\pi}$ cm².

b A circular plate has area 400 cm². Show that its circumference is exactly $40\sqrt{\pi}$

Exercise 4.5f Creating Formulae

1

Pattern 1   Pattern 2   Pattern 3   Pattern 4

  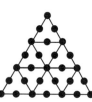

This pattern continues. The sequence of Dot totals is ... 1, 6, 15, 28, ...

a What is the 5th number in this sequence?

b The number of dots, D, in Pattern n is given by $D = an^2 + bn$.
Find the values of a and b. (?)

2 A repair firm's charging schedule is shown.

a Write down the formula for the Total Charge, £C, for a call-out lasting H hours with parts costing £P.

b Change the subject to H.

c Find H if C = 152·5 and P = 40.

> Call-out charge : £50
> Labour charge per hour : £25
> Parts charged at cost

3 A bus averages v km/hr on its 50 km journey into Edinburgh. The average speed for the return journey is normally 10 km/hr faster.

a Write expressions, in terms of v, for the time taken, in hours, for:

  i the journey into Edinburgh

  ii the return journey

b Show that the total time, T hours, for the journey there and back is given by:

$$T = \frac{100(v + 5)}{v(v + 10)}$$

c Show that the average speed, S km/hr, for the journey there and back is given by:

$$S = \frac{v(v + 10)}{v + 5}$$

d Simplify S × T and explain your answer.

4.6 Surds and Indices

Exercise 4.6a Simplifying Surds

1 Express as a surd in its simplest form: (?)

   a $\sqrt{18}$ **b** $\sqrt{75}$ **c** $\sqrt{20}$ **d** $\sqrt{112}$ **e** $\sqrt{128}$

2 Simplify: (?)

   a $3\sqrt{8}$ **b** $5\sqrt{32}$ **c** $2\sqrt{80}$ **d** $4\sqrt{98}$ **e** $3\sqrt{96}$

See pages 39–42 of Leckie & Leckie's Credit Maths Revision Notes

Algebraic Relationships 4

3 Simplify:

a $\sqrt{12} - \sqrt{3}$
b $\sqrt{50} - \sqrt{8}$
c $\sqrt{20} + \sqrt{80}$
d $\sqrt{32} - \sqrt{8} + \sqrt{18}$
e $\sqrt{48} + \sqrt{12} - \sqrt{75}$
f $\sqrt{63} - \sqrt{28} + \sqrt{112}$

4 Multiply out the brackets and express any surds in your answer in their simplest form:

a $\sqrt{3}(\sqrt{6} - \sqrt{3})$
b $\sqrt{5}(\sqrt{5} - \sqrt{10})$
c $\sqrt{2}(\sqrt{10} + \sqrt{14})$
d $(\sqrt{2} + 1)(\sqrt{2} + 3)$
e $(\sqrt{3} - 2)^2$
f $(\sqrt{3} - 2)(\sqrt{3} + 2)$

Exercise 4.6b Rationalising the Denominator

1 Simplify, expressing your answer as a fraction with a rational denominator:

a $\dfrac{\sqrt{12}}{\sqrt{2}}$
b $\dfrac{\sqrt{3}}{\sqrt{12}}$
c $\dfrac{\sqrt{60}}{\sqrt{5}}$
d $\dfrac{\sqrt{24}}{\sqrt{3}}$
e $\dfrac{\sqrt{6}}{\sqrt{12}}$

2 $f(x) = \dfrac{3}{\sqrt{x}}$ and $g(x) = \dfrac{2}{\sqrt{x}}$. Find the exact value of the following. Give your answer as a fraction with a rational denominator.

a $f(8)$
b $g(3)$
c $f(30)$
d $g(10)$
e $f(2) + g(2)$

Exercise 4.6c Working with Indices

1 Express in simplest form:

a $x^6 \times (x^2)^{-1}$
b $\dfrac{a^{5/2} \times a^{-1/2}}{a^2}$
c $\dfrac{2y^4 \times 3y^{-2}}{12y}$
d $\dfrac{(w^{-1})^2}{w^{-1/2} \times w^{-5/2}}$
e $(y^2)^3 \times (y^{-2})^3$
f $\dfrac{3e^{-1/2} \times 4(e^2)^2}{2e^{1/2} \times 2e^2}$

2 Remove brackets and simplify:

a $x^{1/2}(x^{1/2} + x^{-1/2})$
b $a^{-3}(a^6 + a^3)$
c $2y^{-1/2}(\tfrac{1}{2}y^{1/2} - y^{-1/2})$
d $w^{-2}(w^3 - (w^{-1})^{-2})$
e $a^{-1/2}(a + \tfrac{1}{a^{1/2}})$
f $3x^{-3/2}(x^2 + \tfrac{1}{3x^{1/2}})$

3 a Solve:

  i $2^n = 32$
  ii $3^n = 27$
  iii $5^n = \sqrt{5}$
  iv $2^n = 2\sqrt{2}$

 b $f(x) = 2^x$

  i find $f(5)$
  ii If $f(x) = \sqrt{8}$ find x

 c For the formula $V = \dfrac{270}{3^m}$ where $m \geq 0$

  i find V when $m = 2$
  ii find m when $V = \dfrac{10}{9}$
  iii What is the maximum value of V?

See Answers on page 6 of answer booklet

4 Algebraic Relationships

4.7 Solving Quadratic Equations

4.7.1

Exercise 4.7a Quadratic Equations: Solution by Factorising

4.7.2

1. Solve these equations algebraically:

   a $x^2 = 6x$
   b $8y - 2y^2 = 0$
   c $4a = 8a^2$
   d $x^2 + 2x - 15 = 0$
   e $4 - 5x + x^2 = 0$
   f $2w^2 - 11w + 12 = 0$
   g $4y^2 + 4y - 3 = 0$
   h $15x^2 - 11x + 2 = 0$
   i $6x^2 + 5x - 6 = 0$

2. a Two functions are given by $f(x) = x^2 + 3x$ and $g(x) = x + 15$.
     Find the values of x for which $f(x) = g(x)$.

   b Two functions are given by $f(x) = 3x^2 - 3x + 1$ and $g(x) = x - x^2$.
     Find the values of x for which $f(x) - g(x) = 0$.

3.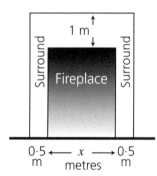

   A fireplace (shaded) is 1 m taller than it is broad. It has a surround 0·5 m wide on each side and 1 m wide at the top as shown. The area of the fireplace is the same as the area of the surround.

   a If the breadth of the fireplace is x m show that $x^2 - x - 2 = 0$.

   b Hence find the breadth of the fireplace and its area.

4.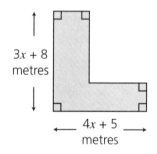

   The diagram shows a garden path which is x metres wide and forms a right-angled L-shape as shown.

   The area of the path is 8 m².

   a Show that $6x^2 + 13x - 8 = 0$.

   b Find the width of the path.

5. The nth triangular number is given by $\frac{1}{2}n(n + 1)$. For example, the 2nd triangular number is given by the substitution $n = 2$:

   $\frac{1}{2} \times 2 \times (2 + 1) = 3$, so 3 is the 2nd triangular number.

   a Show that the statement '28 is the nth triangular number' leads to the equation $n^2 + n - 56 = 0$.

   b Which triangular number is 28? Solve the equation to answer this question.

4.7.3

Exercise 4.7b Quadratic Equations: Solution by Formula

1. Solve these equations, giving your answers correct to 1 decimal place:

   a $x^2 + 4x + 1 = 0$
   b $x^2 + 2x - 1 = 0$
   c $4x^2 - 12x + 3 = 0$
   d $5x^2 - 12x - 8 = 0$
   e $3x^2 - 12x + 11 = 0$
   f $2x^2 + 15x + 16 = 0$

See pages 42–45 of Leckie & Leckie's Credit Maths Revision Notes

Algebraic Relationships 4

2 Solve these equations, giving your answers correct to 2 significant figures:

   a $x^2 + 6x + 4 = 0$ **b** $x^2 - 6x + 3 = 0$ **c** $10x^2 - 7x - 1 = 0$

   d $15x^2 + 2x - 4 = 0$ **e** $3x^2 - 18x + 10 = 0$ **f** $4x^2 - 12x - 15 = 0$

4.8 Variation

Exercise 4.8a Variation: Direct, Inverse and Joint

1 The time, T seconds, for a pendulum to swing varies directly as the square root of its length, l cm.

   a Write down a formula connecting T and l.

   b A pendulum of length 25 cm takes 1 second to swing. Find the time taken for a 1 metre pendulum to swing.

2 The Safe Load, W, of a beam varies directly as the square of its height, h, and inversely as the distance, d, between its supports.

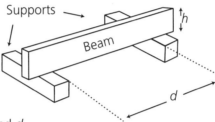

   a Write a formula connecting W, h and d.

   b If $w = 6250$ when $h = 25$ and $d = 20$ calculate W when $h = 30$ and $d = 10$.

3 The Electrical Resistance, R ohms, of a wire varies directly as its length, x metres, and inversely as the square of its diameter, d mm.

   a Write a formula connecting R, x and d.

   b A 2 metre length of wire with 2 mm diameter has resistance 0·5 ohms.

      Find the resistance, in ohms, of a 5 metre length of the same type of wire that has a 4 mm diameter.

Exercise 4.8b Variation: Changes to the Variables

1 The diagram shows a rotating 'Light Display' machine.

The tension, T, in the plastic rod varies directly as the square of the speed, v, of the light and inversely as the length, l, of the rod.

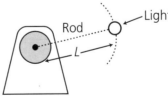

   a Write a formula connecting T, v and l.

   b If the speed and the length are both doubled, what is the effect on the tension?

See Answers on pages 6–7 of answer booklet

4 Algebraic Relationships

2 The pressure, P Newton/m^2, on a circular plug in a water tank varies as the depth, d metres, and as the square of the radius, r metres, of the plug.

   a Write a formula connecting P, d and r.

   b The pressure on the plug in a tank was measured as 3600 Newton/m^2. If the plug is to be replaced by a new plug with half the radius what will be the pressure on the new plug?

3 The volume, V cm^3, of a cylinder varies directly as its height, H cm, and directly as the square of its radius, r cm.

   a Use this information to write a formula connecting V, H and r.

   b If H is doubled and r is halved what effect will this have on the volume?

4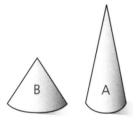

Two cones, A and B, have the same volume. For a fixed volume the height, h cm, of a cone varies inversely as the square of the radius, r cm, of the base.

Cone A has height 18 cm and radius 6 cm. If cone B has height 2 cm, what is its radius?

5 a Explain why y varies directly as x^2.

   b Write down the formula connecting y and x.

   c If x is trebled what effect does this have on y?

x	1	2	3	4
y	4	16	36	64

6 a Explain why q varies inversely as the square root of p.

   b Write down the formula connecting q and p.

   c If p is divided by 4 what is the effect on q?

p	1	4	9	16
q	3	1·5	1	0·75

See pages 45–47 of Leckie & Leckie's Credit Maths Revision Notes

Graphical Relationships 5

5.1 Gradient

Exercise 5.1a Gradient Definition

Give the gradients of the following lines:

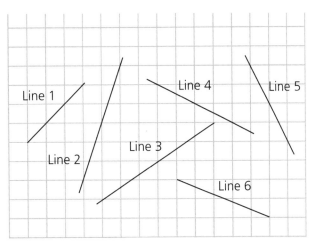

Exercise 5.1b Gradient Formula

1. Calculate the gradient of the line AB for:

   a A (5, 8) and B (2, 2)
   b A (2, 6) and B (1, 9)
   c A (5, −1) and B (1, −3)
   d A (5, 2) and B (−4, −4)
   e A (0, −4) and B (−8, 2)
   f A (−2, −8) and B (−7, −3)
   g A (−4, 0) and B (5, −6)
   h A $\left(\frac{1}{2}, \frac{11}{2}\right)$ and B $\left(-\frac{1}{2}, \frac{3}{2}\right)$
   i A (3, 1) and B $\left(\frac{1}{2}, \frac{3}{2}\right)$ (?)

2. Find the gradient of the line PQ in simplest form: (?)

   a P (m, m^2) and Q (n, n^2)
   b P (3, 9) and Q $(-n, n^2)$
   c P $(a^2, 3a)$ and Q $(b^2, 3b)$
   d P (c, ac^2) and Q (d, ad^2)

5.2 Linear Graphs

Exercise 5.2a Graphs from Equations

1. Two variables x and y are connected by the relationship $y = mx + c$. (?)

   Draw a sketch of a possible graph for each of these cases:

   a $m > 0$ and $c > 0$
   b $m < 0$ and $c > 0$
   c $m > 0$ and $c < 0$
   d $m < 0$ and $c < 0$
   e $c = 0$
   f $m = 0$

2. Sketch graphs for the following:

   a $y = 2x + 1$
   b $y = -x + 3$
   c $y = x - 2$
   d $y = \frac{1}{2}x - 1$
   e $y = -\frac{2}{3}x + 4$
   f $2y + x = 12$

5 Graphical Relationships

5.2.2

Exercise 5.2b Equations from Graphs

1. Find the equations of these straight lines:

   a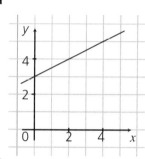
   b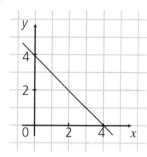
   c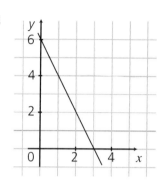
   d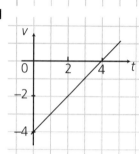
   e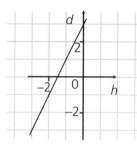
   f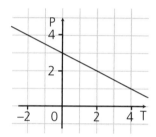

2. For each diagram: **i** find the gradient of AB **ii** write down the equation of AB

   a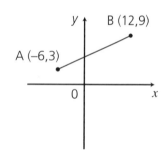
   b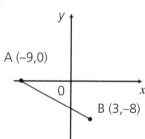
   c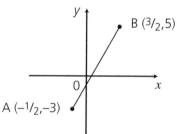

   a AB cuts the y-axis at (0, 5)
   b AB cuts the y-axis at (0, −6)
   c AB cuts the y-axis at (0, −1)

3. The graph represents the distance (d km) to Glasgow against time (t hours) from the start of a train's journey.

   a Find the equation of the line in terms of d and t.
   b How long does the train take to travel 100 km?

   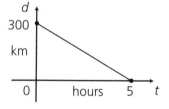

4.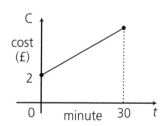

   The graph shows how the cost, £C, of an international phone call increases with the time, t minutes, connected.

   The 30 minute call cost £5. There was a £2 connection fee.

   a Find the equation of the straight line in terms of C and t.
   b A similarly charged call cost £3·90 in total. How long was the call?

See pages 48–50 of Leckie & Leckie's Credit Maths Revision Notes

Graphical Relationships 5

5 Find the equation of each line in terms of the two given variables:

a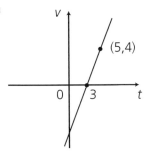

b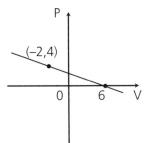

c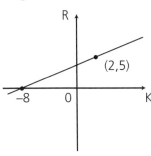

Exercise 5.2c What's the Point?

1 Which of the points A(−3, 1), B(2, 4) and C(3, 2) lies on each of these lines:

   a $5y - 3x = 14$
   b $y + 2x = 8$
   c $6y - x = 9$

2 a $A(2k, k)$ lies on $y = 2x + 3$. Find the coordinates of A.

   b $P(a, a)$ lies on $y = -\frac{1}{2}x - 3$. Find the coordinates of P.

   c $Q(m, m^2)$ lies on $y = 6 - x$. Given that $m > 0$ find the coordinates of Q.

5.3 Solving Problems Graphically

Exercise 5.3a Graphs: Simultaneous Equations

1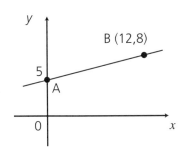

The diagram shows a railway line running between two stations A and B.

   a Show that the equation of line AB is $4y - x = 20$.

   b Another line, not shown, has equation $3y + 4x = 34$. This line meets the line AB at station C. Calculate the coordinates of C.

2 The diagram shows two motorways A and B with their intersection at point R. Two petrol stations are shown: P(−2, 6) on motorway A and Q(4, 7) on motorway B. Motorway B crosses the x-axis at $(-\frac{2}{3}, 0)$ and motorway A crosses the y-axis at (0, 5).

From the information given, calculate the coordinates of point R.

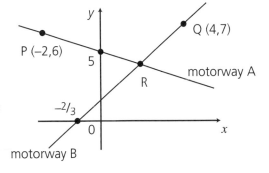

5 Graphical Relationships

Exercise 5.3b Graphs: Iteration

1. The equation $x^3 - 3x^2 + 1 = 0$ has a root between 0 and 1. Use iteration to find the value of the root correct to 1 decimal place. Show all your working clearly.

2. There is a root of the equation $x^4 + 4x - 2 = 0$ somewhere between -1 and -2. Use iteration to determine the value of this root correct to 1 decimal place showing your working clearly.

3. The equation $1 + 3x - x^3 = 0$ has a total of three roots between -2 and 2. Find each of these roots correct to 1 decimal place.

5.4 Quadratic Graphs

Exercise 5.4a Quadratic Graphs

1. For each diagram use the given information to find the value of k:

   a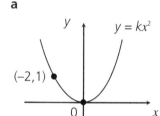
   b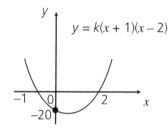
   c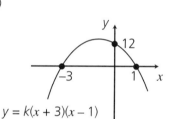

2. For each of the following graphs:
   i   write down the coordinates of A
   ii find the coordinates of B and C
   iii calculate the coordinates of the minimum of maximum point D.

   a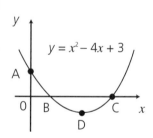
   b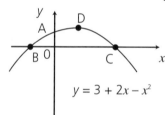
   c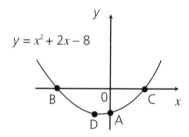

   d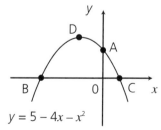
   e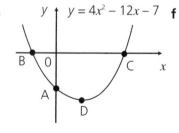
   f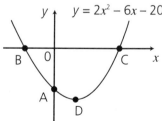

See pages 50–52 of Leckie & Leckie's Credit Maths Revision Notes

Graphical Relationships 5

3 The diagram shows part of a quadratic graph with equation:
$y = k(x - a)(x - b)$.

The graph cuts the x-axis at $(-3, 0)$ and $(2, 0)$ and it cuts the y-axis at $(0, -24)$.

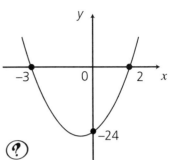

a Write down the values of a and b.

b Calculate the value of k.

c Find the coordinates of the minimum turning point of the graph.

Exercise 5.4b Quadratic Graphs: Max/Min Problem Solving

1 To answer the question, in each case you should find where the graph cuts the x-axis:

a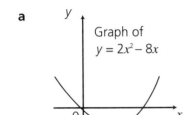

Find the least positive value of x for which $2x^2 - 8x \geqslant 0$

b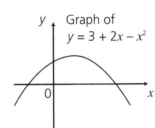

Find the maximum value of x for which $3 + 2x - x^2 \geqslant 0$

c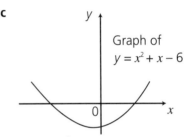

Find the least positive value of x for which $x^2 + x - 6 \geqslant 0$

2 This 30 cm × 40 cm painting is being framed. The frame has width x cm.

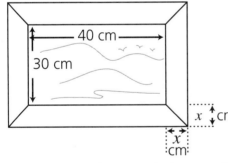

a Show that the area, A cm², of the painting *and* the frame is given by

$A = 4x^2 + 140x + 1200$.

b It is recommended that the area of the frame should be at least $\frac{2}{3}$ of the area of the painting.

Show that $4(x - 5)(x + 40) \geqslant 0$.

c Find the minimum possible width for the frame. Show clearly your reasoning.

3 A pencil holder is made from a strip of metal with dimensions 10 cm × 40 cm (step 1). The ends are folded up with the fold at one end three times the length of the fold at the other end (step 2). Plastic sides are then added (step 3).

a Show that the volume, V cm³, of this prism-shaped holder is given by $V = 800x - 80x^2$.

b Calculate the maximum possible volume for this design.

c If the volume has to be at least 1280 cm³

  i show that $80(x^2 - 10x + 16) \leqslant 0$

  ii find the minimum possible value for x.

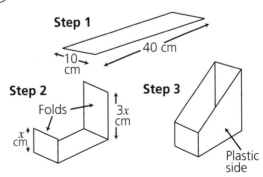

See Answers on page 8 of answer booklet

5 Graphical Relationships

4 A fenced enclosure is planned at the back of a 20 m × 25 m house. 100 m of fencing will be used for the rectangular enclosure which will include the back wall of the house as part of one side as shown in the diagram.

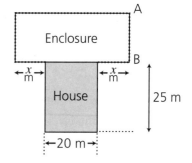

a Show that AB = (40 − 2x) m.

b Show that the area, E m^2, is given by E = 800 + 40x − 4x^2.

c Find, algebraically, the maximum possible area for the enclosure.

d If the enclosure has to be at least 75% larger than the area of the house

 i show that $4x^2 - 40x + 75 \leq 0$

 ii hence find the minimum and maximum possible values for x.

5.5 Other Types of Graphs

Exercise 5.5

1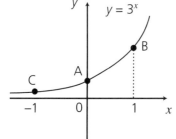

Write down the coordinates of A, B and C.

2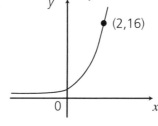

Calculate a.

3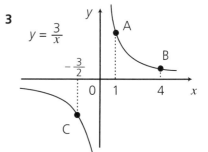

Find the coordinates of A, B and C.

Graphs, Statistics and Probability 6

6.1 Displaying Data

Exercise 6.1a Constructing Pie Charts

Draw a pie chart to illustrate the results of each survey:

1 Favourite Colour

Red	14
Green	2
Blue	3
Yellow	1

2 Favourite Flavour

Salt & Vinegar	25
Plain	12
Beef	8
Cheese & Onion	15

3 Travel to work

Walking	30
Train	69
Bus	55
Taxi	26

Exercise 6.1b Quartiles and Boxplots

1 Find the median, lower quartile and upper quartile for the following sets of data. Draw a boxplot to illustrate the data in each case.

a 9, 3, 4, 6, 2, 7, 6

b 1, 7, 11, 10, 3, 5

c 12, 3, 8, 9, 11, 4, 1, 8, 8

d 19, 13, 15, 22, 11, 17, 16, 23

2 The average monthly rainfall, in mm, is shown for a village:

Month :	Jan	Feb	Mar	Apr	May	Jun	Jul	Aug	Sep	Oct	Nov	Dec
Rainfall (mm) :	2	3	3	5	4	5	7	8	4	3	1	1

Draw a suitable statistical diagram to illustrate the median and the quartiles of this data.

3 A survey of the lifetimes, in months, of two different types of light bulbs gave the following statistics:

	minimum	maximum	Lower Quartile	Median	Upper Quartile
Type A	1	34	15	21	25
Type B	3	31	9	24	26

a Draw an appropriate statistical diagram to illustrate these two sets of statistics.

b Which type of bulb is more reliable? Give a reason for your answer.

4 A random sample of 10 children at a school were asked how many hours a week they spent watching television. The results were:

12, 6, 7, 16, 10, 22, 11, 8, 11, 6

The parents of these same children were then asked the same question. Here are the results:

11, 3, 3, 12, 7, 15, 11, 5, 10, 2

Draw an appropriate statistical diagram to compare these two sets of data.

Exercise 6.1c Scattergraphs/Lines of Best Fit

1 The diagram shows a scattergraph comparing Physics score, p%, with Maths score, m%, gained by one particular class of students at a school. Student A scored 0% in Maths and 20% in Physics. Student B scored 90% in Maths and 80% in Physics.

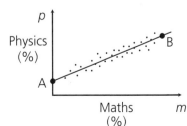

a Find the equation of line of best fit AB in terms of p and m.

b Iain was absent for the Maths Test but scored 56% in the Physics Test. Use the equation to estimate his Maths score.

6 Graphs, Statistics and Probability

2 Crops were sprayed with a chemical and the percentage of plants still affected by mildew were recorded. Different crops were sprayed with different concentrations of the chemical. The results of the trial are shown in the scattergraph with the line of best fit drawn in.

a Describe the correlation between the percentage infected and the concentration.

b Crop A was not treated and 95% of the plants were infected. Crop B was sprayed with a concentration of 2000 units/hectare with a result that only 15% of the plants were infected. Find the equation of AB, the line of best fit, in terms of P and U.

c If the acceptable infection level is 24% maximum, use the equation to recommend a minimum concentration level for spraying crops.

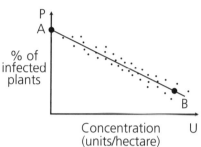

6.2 Statistics

Exercise 6.2a Quartiles and Semi-interquartile Range

1 Calculate the median and semi-interquartile range for each of these data sets:

a 2, 8, 3, 5, 3, 4, 4

b 2·8, 3·5, 1·7, 1·8, 2·9, 3·0

c 14, 27, 26, 29, 15, 21, 21, 20, 30

d 0·8, 1·5, 0·2, 0·1, 2·0, 1·7, 0·9, 1·3

2 The heights, in cm, of a random sample of seedlings were measured. This boxplot summarises the result:

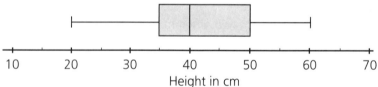

a What was the range of heights?

b Calculate the semi-interquartile range for the heights.

c What % of the seedlings were:

i less than 35 cm

ii greater than 40 cm

iii less than 50 cm?

Exercise 6.2b Frequency Tables

For each table of data calculate:

a the mode
b the mean
c the median
d the upper and lower quartiles
e the range
f the semi-interquartile range

1 The family size (number of children) for each of 30 students was recorded:

Family size :	1	2	3	4	5	6
Frequency :	5	11	8	3	2	1

2 The time taken by 45 workers to complete a task were recorded to the nearest minute:

Number of minutes :	8	9	10	11	12	13	14
Frequency :	11	4	16	8	2	0	4

See pages 58–62 of Leckie & Leckie's Credit Maths Revision Notes

Graphs, Statistics and Probability 6

Exercise 6.2c Standard Deviation

1 A random sample of matchboxes is checked. The total number of matches in the box are:

62, 62, 61, 57, 63

  a Calculate the mean and standard deviation for the sample.

  b 'On average 60 matches'. Comment on this statement printed on each box.

2 Alisdair is comparing download prices, in pence, for music files. He visited six different music sites. Prices per track are:

62, 59, 58, 63, 64, 60

  a Find the mean price for a download.

  b Calculate the standard deviation for these prices to 1 decimal place.

  c Equivalent prices at retail shops have the same mean but with a standard deviation of 4·1. Make a valid comparison between download prices and retail shop prices.

3 Buying prices, in £, for lambs at a market town auction are:

123, 129, 120, 119, 146, 120, 125, 104, 118, 121

This was a random sample of 10 prices in July.

  a Calculate the mean price of a lamb.

  b Calculate the standard deviation for these prices to 1 decimal place.

  c Make two valid comparisons between these market town auction prices and a rural auction in July where the mean price was £118 with the standard deviation being 2·5.

6.3 Probability

Exercise 6.3 Probability

1 A bag contains 20 marbles, 5 each of red, blue, white and black. A marble is picked at random. What is the probability that it is:

  a white   **b** red or blue   **c** not black   **d** yellow

2 A survey was carried out on 1000 randomly selected sheep on Mull to determine their age profile. Here are the results:

Age in Years :	under 1	between 1 and 2	between 2 and 3	between 3 and 4	between 4 and 5	over 5
female :	300	110	80	100	60	30
male :	300	0	0	5	10	5

  a What is the probability that a randomly chosen sheep on Mull is:

    i younger than 2 years   **ii** male   **iii** female older than 3 years

  b In a 2400 sheep flock how many sheep would you expect to be:

    i female   **ii** under 1 year old   **iii** male older than 3 years

See Answers on page 9 of answer booklet

Exam Structure and Formulae

The current structure of the SQA Mathematics Standard Grade Credit Level Exam is as follows:

Paper 1 (Non-Calculator)   Time: 55 min
Paper 2 (Calculator allowed)   Time: 80 min

You will be reminded that you should answer as many questions as you can and that to gain full credit in a question your solution must contain appropriate working. Square-ruled paper will be provided if you need this for a particular question.

The following Formulae List is given to you in the exam:

The roots of $ax^2 + bx + c = 0$ are $x = \dfrac{-b \pm \sqrt{(b^2 - 4ac)}}{2a}$

Sine rule: $\dfrac{a}{\sin A} = \dfrac{b}{\sin B} = \dfrac{c}{\sin C}$

Cosine Rule: $a^2 = b^2 + c^2 - 2bc \cos A$ or $\cos A = \dfrac{b^2 + c^2 - a^2}{2bc}$

Area of a triangle: Area $= \tfrac{1}{2}ab \sin C$

Standard Deviation: $s = \sqrt{\dfrac{\Sigma(x - \bar{x})^2}{n - 1}} = \sqrt{\dfrac{\Sigma x^2 - (\Sigma x)^2 / n}{n - 1}}$, where n is the sample sign

Practice Exams

Practice Exam A

Paper 1 (Non-Calculator)   **Time allowed: 55 minutes**

1. Evaluate $3·02 + 0·1 \times 2$

2. Evaluate $\frac{3}{7}(1\frac{2}{3} - \frac{1}{2})$

3. Simplify $\frac{5}{x+1} - \frac{2}{x}$

4. $g(x) = x^2 - 2x$. Evaluate $g(-3)$

5. The number of paper clips in a random sample of 10 boxes is:

    102, 95, 94, 99, 94, 105, 97, 101, 97, 101

    Draw a suitable statistical diagram to illustrate the median and the quartiles of this data.

6. Luke receives a $7\frac{1}{2}\%$ increase in salary and he now earns £8600 per year as a trainee chef. What was his salary before the increase?

7. A dodecahedral die has the numbers 1 to 12 on the faces.
   A normal cubical die has the numbers 1 to 6 on its faces.

   With which die do you have a better chance of rolling a prime number?

   Show clearly all your working.

8. In this numberbrick pattern, the number on each brick is the sum of the numbers of the two bricks below it.

   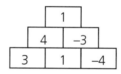

   a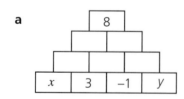

   By completing this numberbrick pattern show that $x + y = 2$.

   b

   The same x and y are used in this numberbrick pattern.

   Find another equation in x and y.

   c Find the values of x and y.

Practice Exam A

9 Part of the graph of $y = 4 \sin 2x°$ is shown below.

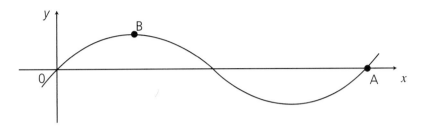

Write down the coordinates of the points A and B.

10 Part of the graph of $y = kx(x - 5)$ is shown. The point A(2, 18) lies on the graph. Calculate the value of k.

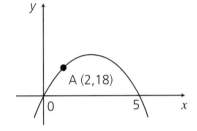

11 a Simplify $\sqrt{125} - \sqrt{45}$

  b Evaluate $27^{-2/3}$

12 This equilateral triangle has a height of exactly $3\sqrt{3}$ cm. Calculate the length of its sides.

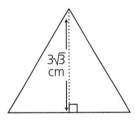

Practice Exam A

Paper 2 (Calculator allowed) Time allowed: 80 minutes

1. The volume, V, of a sphere is given by $V = \frac{4}{3}\pi r^3$ where r is the radius of the sphere. Taking the radius of the Earth to be $6\cdot 4 \times 10^3$ km, calculate its volume in km³. Give your answer in scientific notation to 2 significant figures.

2. A scattergraph is shown comparing weight, w kg, with age, m months for a particular group of babies. Baby A when born weighed 3 kg and baby B at 16 months old weighs 5 kg.

   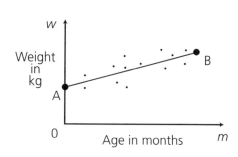

   a Find the equation of the line of best fit AB in terms of w and m.

   b Use the equation to estimate the age of a 3·5 kg baby from this group.

3. The numbers of absentees at a school during one week were as follows:

   $$32,\ 29,\ 33,\ 33,\ 38$$

   Calculate the mean and standard deviation for this data.

4. The value of Mr and Mrs Wilson's house is increasing at the rate of 2·5% each month. At the start of March their house is valued at £185 000. Calculate its value by the start of June. Give your answer to the nearest £1000.

5.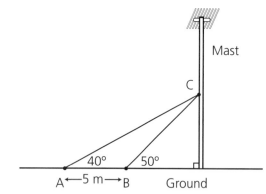

   Two support cables are attached to a mobile phone mast at point C as shown in the diagram. The cables make angles of 40° and 50° to the ground at points A and B which are 5 m apart.

   Calculate, to the nearest cm, the height of point C above the ground.

6. The volume, V cm³, of a fixed amount of gas varies at the temperature, T kelvins, and inversely as the pressure, p mmHg.

   a Write down a formula for V in terms of T and p.

   b At a temperature of 450 kelvins and a pressure of 750 mmHg the gas has a volume of 510 cm³. If the pressure is now increased by 250 mmHg with the volume remaining the same, calculate the resulting temperature of the gas.

7 Measurements were taken from a radar display showing the position of three boats A, B and C. From A, boat B is 12 km away and boat C is 5 km away. The angle between their directions is 125° as shown on the diagram.

Calculate, to 3 significant figures, the distance of boat B from boat C.

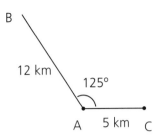

8 The diagram shows the largest cross-section of a spherical glass paperweight with radius 3·4 cm. The base of the paperweight has diameter 3·1 cm.

Calculate the height, h cm, of the paperweight.

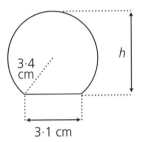

9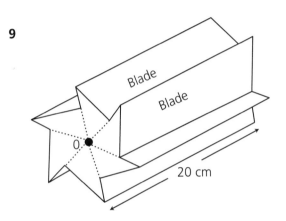

A rotor fan in the shape of a prism has six identical blades that fit around the central point 0. Each blade has a uniform cross-section with 7 cm and 5 cm sides that form a triangle meeting at 0.

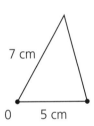

The rotor is 20 cm in length. Calculate the volume of the rotor.

10 Solve algebraically the equation

$$2 \tan x° + 2 = \sin 80° \qquad 0 \leqslant x < 360$$

11 A symmetrical lawn consists of two identical rectangles as shown. The breadth of each rectangle is x metres and has length twice its breadth. A path, 1 m wide surrounds the lawn on all sides.

a If the total area of the lawn is 20 m² more than the area of the path, show that

$$4x^2 - 10x - 24 = 0$$

b Hence find the dimensions of each of the rectangular pieces of the lawn.

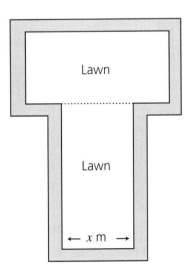

Practice Exam B

Paper 1 (Non-Calculator)   **Time allowed: 55 minutes**

1. Evaluate $\dfrac{10{\cdot}4}{9{\cdot}35 - 6{\cdot}85}$

2. Evaluate $1\tfrac{1}{4} \div 3\tfrac{1}{3}$

3. Solve the inequality: $3(5 - x) \leqslant 5x - 1$

4. Given that $f(x) = 3x - 2x^2$ evaluate $f(-4)$

5.
   a. Factorise $50A^2 - 32$
   b. Hence simplify $\dfrac{5A + 4}{50A^2 - 32}$

6. The diagram shows the line AB where A(0, –5) and B(15, 5)

   a. Find the equation of line AB.
   b. Calculate the x-coordinate of the point C where the line AB cuts the x-axis.

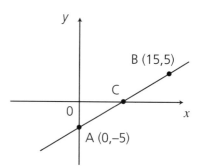

7. Maureen and Ken are buying bread and rolls.

   a. Maureen buys 5 loaves and half a dozen rolls at a cost of £5·81. Write down an algebraic equation to illustrate this.
   b. Ken's bill comes to £3·59. He bought 3 loaves and 4 rolls. Write down an algebraic equation to illustrate this.
   c. Find the total cost of a dozen rolls and 4 loaves.

8. An octohedral die has the numbers 1 to 8 on its eight faces.

   The die is rolled.

   a. What is the probability that the top face shows a square number?
   b. What is the probability that the top face does not show a multiple of 3?

9. A batch of seedlings had their height recorded in cm. The results are shown in the boxplot.

   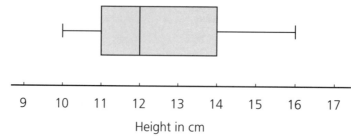

   Height in cm

   a. State the upper quartile.
   b. Calculate the semi-interquartile range.
   c. What percentage of the seedlings were taller than 11 cm?

Practice Exam B

10 An orchestra has string players, wind players and percussion players. The ratio of percussion to wind to string is 1:3:5.

   a With 25 string players in the orchestra how many wind players would there be?

   b If the combined wind and percussion sections have a total of 24 players, how many string players would there be?

11 The sequence 1, 5, 6, 11, 17, … has the rule that after the first two terms each term is the sum of the two previous terms. The sequence $m, n, m + n, …$ has the same rule.

   a Show that t_5, the 5th term is $3m + 5n$.

   b Show that $t_5 = 3t_3 + 2t_2$.

12 a Evaluate $25^{-3/2}$

   b Simplify $1 - \frac{\sqrt{3}}{\sqrt{6}}$ writing your answer as a single fraction with a rational denominator.

13 The diagram shows the end of a house along with an extension. The house end is a rectangle with dimensions $2x$ m × $3x$ m surmounted by an isosceles triangle with height x metres.

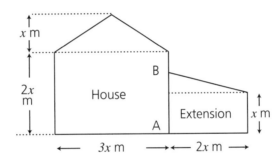

The extension is a $2x$ m × x m rectangle surmounted by a right-angled triangle attached to the side of the house at point B at the top and point A at the bottom.

The area of the house end is three times the area of the extension end. Find, in terms of x, the length of AB.

Practice Exam B

Paper 2 (Calculator allowed)   **Time allowed: 80 minutes**

1. A road haulage firm has a yearly running cost of £375 000. The financial department is working on a 3 year plan and makes the assumption that running costs will increase at the rate of 3·8% per year. Calculate, to the nearest £1000, their estimate for the running cost per year in 3 years' time.

2. The number of hours of sunshine was recorded daily in Stonehaven over the course of a week. The results were as follows:

Sun	Mon	Tue	Wed	Thu	Fri	Sat
2·4	3·4	4·9	4·0	3·8	3·8	3·6

   a Find the mean number of hours of sunshine that week.

   b Find the standard deviation of the data.

   c Kenmore, with the same mean number of hours of sunshine that week, had a standard deviation of 1·3. Make one valid comparison between the two places.

3.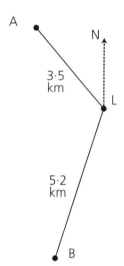

   From a lighthouse L, Rock A has a bearing of 320° and Rock B has a bearing of 195°.

   Rock A is 3·5 km from L and Rock B is 5·2 km from L.

   Calculate the distance in km, to 1 decimal place, between the two rocks. (Do not use a scale drawing.)

4. The volume, V cm³, of the cone shown is given by
   $$V = \tfrac{1}{3}\pi r^2 h$$
   where r cm is the radius of the base circle and h cm is the height.

   The cone has a height of 18 cm and base diameter of 12 cm.

   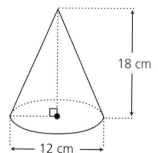

   a Calculate the volume of the cone.

   b A cube has the same volume as this cone.

   Calculate the length of its edges to the nearest mm.

5. The formula $F = \tfrac{9C}{5} + 32$ is used to convert temperatures from the Celsius scale to the Fahrenheit scale. For example, when C = 5, F = 41 so a temperature of 5°C is converted to 41°F.

   Use the formula to convert a temperature of 66·2°F to Celsius.